Basics of Cloud Computing

and

I0504889

Artificial Intelligence

Isaac Odun-Ayo Ph.D.

ISBN: 9798394326653

DEDICATION

Dedicated to my lovely children Funmilola, David, Seyi, Titilayo, Ope and Moses

CONTENTS

ACKNOWLEDGMENTS

Thanks be to God, my colleagues and family

i

1 INTRODUCTION TO CLOUD COMPUTING AND ARTIFICIAL INTELLIGENCE

Cloud computing and artificial intelligence are two rapidly emerging technologies that are reshaping our lives and workplaces. This chapter will provide an overview of these two technologies, including definitions, historical context, and evolution.

1.1 Cloud Computing Explained

Cloud computing is a methodology for offering computer resources and services such as servers, storage, databases, networking, software, and analytics over the internet. These resources and services are available on a pay-as-you-go basis, allowing consumers to adjust their usage as needed. Cloud computing is frequently referred to as the "utility computing" of the twenty-first century since it provides computing resources and services on demand, similar to energy or water utilities.

There are various cloud computing models available, including public cloud, private cloud, and hybrid cloud. Third-party providers such as Amazon Web Services (AWS), Microsoft Azure, and Google Cloud provide public cloud services. Private cloud services are offered by an organization's own data center or a third-party vendor with dedicated resources. Hybrid cloud mixes public and private cloud services, allowing enterprises to reap the benefits of both.

1.2 Artificial Intelligence Definition

The simulation of human intelligence processes by machines, particularly computer systems, is known as artificial intelligence (AI). AI technology can be utilized to accomplish activities that would normally need human intellect, such as speech recognition, image recognition, and decision making. Artificial intelligence (AI) is a vast field that includes various subfields such as machine learning, deep learning, natural language processing, computer vision, and robotics.

Machine learning is a subset of AI in which computers are trained to make predictions or judgments based on data without being explicitly programmed. Deep learning is a kind of machine learning in which neural networks with numerous layers are trained to recognize patterns and make choices. Natural language processing refers to computers' ability to understand and interpret human language, whereas computer vision refers to computers' ability to analyze visual data such as photographs and videos.

1.3 Cloud Computing and Artificial Intelligence Evolution

Cloud computing and artificial intelligence have a long history of development and advancement. With the introduction of virtualization technology in the 1990s, which allowed several virtual machines to run on a single physical machine, the first cloud computing services arose. Cloud computing became popular in the early 2000s as customers sought access to computing resources and services from anywhere, at any time.

In the early days of AI, researchers concentrated on rule-based systems that made choices using if-then expressions. However, advances in machine learning and deep learning algorithms in the 2010s enabled computers to learn from data and recognize patterns. As a result, AI-powered products and services such as voice assistants, recommendation systems, and self-driving automobiles have been developed.

Cloud computing and AI are merging today, with cloud providers providing AI-as-a-Service platforms that allow customers to design, train, and deploy AI models on the cloud. This convergence is driving the creation of new applications and services that are altering industries and reshaping how people live and work.

1.4 Cloud Computing and Artificial Intelligence Have Many Advantages

The integration of cloud computing and artificial intelligence provides various advantages for businesses, including:

1. **Scalability**: Cloud computing enables enterprises to effortlessly scale up or down their computing resources and services based on their needs. AI models can also be scaled up or down based on the amount of training data available.

2. **Flexibility**: Cloud computing and AI enable firms to test new technologies and processes without investing heavily in infrastructure or expertise.

3. **Cost savings**: By providing computer resources and services on a pay-

as-you-go basis, cloud computing and AI can assist enterprises in reducing costs. Large capital investments in on-premise infrastructure are no longer required.

4. **Speed:** Cloud computing and artificial intelligence enable enterprises to rapidly install and scale new apps and services, lowering time-to-market and increasing agility.

5. **Cloud computing and artificial intelligence** (AI) are enabling enterprises to construct previously unachievable apps and services, resulting in new business models and income streams.

1.5 Cloud Computing and Artificial Intelligence Challenges

While the convergence of cloud computing and artificial intelligence provides several benefits, organizations must also consider the following challenges:

1. Cloud computing and artificial intelligence pose additional security threats, such as data breaches, illegal access, and hostile attacks.

2. Data privacy: Cloud computing and AI necessitate vast volumes of data to be collected, stored, and processed, which creates issues about data privacy and compliance with rules such as GDPR and CCPA.

3. Cloud computing and AI necessitate specific skills, which might be challenging to find and keep.

4. Complexity: Cloud computing and artificial intelligence (AI) are complicated technologies that necessitate extensive knowledge to conceive, develop, and deploy.

2 CLOUD COMPUTING FUNDAMENTALS

Cloud computing has become an indispensable component of modern technology, with organizations and individuals alike utilizing its services for data storage, application hosting, and other computing chores. Cloud computing principles serve as the foundation for cloud-based services, allowing businesses to scale up and down as needed while also offering reliable access to data and applications. In this chapter, we will look at the fundamentals of cloud computing architecture, several types of cloud computing models, and the pros and drawbacks of cloud computing.

2.1 The Architecture of Cloud Computing

Several key components make up cloud computing architecture, including:

1. **Infrastructure-as-a-Service (IaaS):** This cloud computing layer offers virtualized computer resources such as servers, storage, and networking. It gives organizations more control over their infrastructure, allowing for greater customization and flexibility.

2. **Platform-as-a-Service (PaaS):** This cloud computing layer provides a platform for enterprises to design and deploy applications without having to manage the underlying infrastructure. It covers development, testing, and deployment tools and services.

3. **Software as a Service (SaaS):** This cloud computing layer delivers entire apps that may be accessed via a web browser or API. It enables enterprises to employ software programs without the need for on-premises installation and management.

2.2 Models of Cloud Computing

The three major cloud computing models are public, private, and hybrid.

1. Public cloud services are open to anybody with access to the internet and are maintained by third-party providers. They provide a pay-as-you-go model, which allows organizations to begin adopting cloud computing without making any upfront investments in hardware or infrastructure.

2. Private Cloud is a single entity owns and operates private cloud services, which are either hosted on-premises or in a data center. They provide greater control and customization over the infrastructure, but need a larger initial investment.

3. Hybrid cloud services combine the benefits of both public and private clouds, allowing businesses to use a mix of public and private cloud services. Businesses can benefit from the scalability and flexibility of public cloud services while keeping essential data and apps secure in a private cloud.

2.3 The Advantages of Cloud Computing

Cloud computing provides various advantages to enterprises, including:

1. **Scalability**: Cloud computing enables businesses to effortlessly scale up or down their computing resources based on their demands. This allows firms to absorb seasonal traffic spikes or accommodate growth without having to invest in new hardware.

2. **Savings on On-Premises Infrastructure**: Cloud computing eliminates the need for big upfront expenditures in on-premises infrastructure. Instead, firms pay only for the resources they use, making cost management and resource allocation more efficient.

3. Cloud computing provides high availability and uptime, ensuring that programs and data are constantly available. This is accomplished by incorporating redundancy and failover methods into the cloud architecture.

4. Security hazards associated with cloud computing include data breaches, illegal access, and malicious attacks. To safeguard data and applications, cloud service providers have adopted a variety of security measures, including encryption, firewalls, and access controls.

2.4 Cloud Computing's Difficulties

While cloud computing has many advantages, it also has some drawbacks that businesses must consider, including:

1. Security hazards associated with cloud computing include data breaches, illegal access, and malicious attacks.

2. Compliance: Companies must verify that their cloud computing activities are in accordance with regulations such as GDPR, HIPAA, and PCI DSS.

3. Data privacy: Businesses must ensure that their cloud computing operations safeguard customer data privacy.

4. To avoid vendor lock-in, businesses must carefully analyze the terms and conditions of their cloud computing service providers.

Businesses can make informed decisions about which cloud computing services to utilize, how to control expenses, and how to optimize their cloud infrastructure for optimal efficiency and security by understanding the foundations of cloud computing.

3 FUNDAMENTALS OF ARTIFICIAL INTELLIGENCE

1. **Artificial intelligence** (AI) is a subfield of computer science that focuses on developing machines capable of doing activities that normally require human intelligence, such as visual perception, speech recognition, decision-making, and natural language processing. Artificial intelligence (AI) is a vast field that includes various subfields such as machine learning, natural language processing, robotics, computer vision, and expert systems.

2. **Machine learning** is an AI subfield that focuses on creating systems that can learn and improve based on data. Machine learning is classified into three types: supervised learning, unsupervised learning, and reinforcement learning.

3. **Supervised learning** is a sort of machine learning in which the computer is trained on a labeled dataset, which means that the expected output is provided for each input. Based on the training data, the computer then learns to map inputs to outputs. Image recognition, speech recognition, and language translation are examples of tasks that require supervised learning.

4. **Unsupervised learning** is a subset of machine learning in which the computer is trained on an unlabeled dataset with no predefined output for each input. After that, the machine learns to look for patterns and structure in the data. Clustering, anomaly detection, and dimensionality reduction are examples of unsupervised learning problems.

5. **Reinforcement learning** is a sort of machine learning in which the machine learns by interaction with its environment and feedback in

the form of incentives or punishments. Following that, the machine learns to take actions that maximize the cumulative reward over time. Reinforcement learning is utilized in tasks such as gaming, robotics, and self-driving cars.

AI offers a wide range of applications in industries such as healthcare, banking, manufacturing, and transportation. Here are some examples of AI applications:

1. AI can be applied in healthcare for activities such as disease detection, medication discovery, and medical imaging analysis.

2. Finance: Artificial intelligence (AI) can be utilized for tasks such as fraud detection, risk management, and algorithmic trading.

3. Manufacturing: Artificial intelligence (AI) can be utilized for activities including predictive maintenance, quality control, and supply chain optimization.

4. Transportation: AI may be used to do activities such as self-driving, traffic control, and logistics optimization.

AI has also been used to create virtual assistants, Chatbots, and recommender systems, which deliver personalized recommendations to users based on their tastes and behavior. As AI technology advances, it has the ability to alter business operations and transform entire industries. However, in order to make informed decisions about how to leverage this technology for maximum benefit, businesses must first understand the fundamentals of AI and its applications.

4 AI CLOUD COMPUTING INFRASTRUCTURE

Because of its scalability, flexibility, and low cost, cloud computing is a perfect platform for AI workloads. Not all cloud computing infrastructures, however, are appropriate for AI workloads. To manage the processing and storage requirements of massive datasets and complicated algorithms, AI workloads necessitate specialized infrastructure. In this chapter, we'll look at the infrastructure needs for AI workloads in the cloud, as well as the various cloud computing platforms that can support them.

4.1 Infrastructure Needs for AI Workloads

AI workloads have distinct infrastructure requirements than traditional computer workloads. High-performance computing, distributed computing, and large-scale storage and data management skills are required for AI applications. Some of the most important infrastructure needs for AI workloads are as follows:

1. AI workloads necessitate high-performance computing in order to execute complicated computations on massive datasets. GPUs and TPUs, which are geared for parallel processing and can do computations quicker than standard CPUs, are examples of high-performance computing hardware.

2. AI workloads frequently demand distributed computing to address the processing and storage requirements of huge datasets. Distributed computing entails breaking down a burden into smaller tasks and distributing them for processing over numerous servers or nodes.

3. AI workloads necessitate large-scale storage and data management skills in order to handle the storage and retrieval of massive datasets.

4.2 Cloud High-Performance Computing and Distributed Computing

Cloud computing offers a variety of high-performance computing and distributed computing technologies that are well-suited to AI applications. The following are some of the most important cloud computing technologies for AI workloads:

1. Cloud companies offer Graphics Processing Unit (GPU) and Tensor Processing Unit (TPU) instances, which are geared for high-performance computing and can do computations quicker than standard CPUs. These instances are suited for artificial intelligence workloads that demand parallel processing and large-scale computations.

2. Serverless computing is a cloud computing architecture in which the cloud provider manages the infrastructure needed to operate an application. Serverless computing frees developers from the burden of managing infrastructure, making it perfect for AI workloads that demand rapid development and deployment.

3. Containers and container orchestration: Containers are self-contained, lightweight environments that can be used to bundle and deliver software. Container orchestration systems like Kubernetes enable the management and scalability of containerized applications, making them perfect for AI workloads requiring distributed computing.

4.3 AI Data Storage and Management in the Cloud

Several storage and data management technologies are available in the cloud that are ideal for AI workloads. The following are some of the most important cloud computing technologies for AI workloads:

1. Object storage is a cloud storage strategy in which data is stored as objects rather than files or blocks. Object storage is appropriate for AI workloads that demand huge dataset storage and retrieval.

2. Data lakes are centralized repositories that store massive amounts of raw data in its original format. Data lakes are perfect for AI tasks that need huge dataset processing and the building of machine learning models.

3. Data warehouses are cloud-based databases that are designed for querying and analysis. Data warehouses are suited for AI tasks that necessitate data processing and visualization.

Finally, because of its scalability, flexibility, and cost-effectiveness, cloud computing is an ideal infrastructure for AI workloads. AI workloads have distinct infrastructure requirements than traditional computer workloads. Cloud computing offers a variety of high-performance computing and distributed computing technologies, as well as storage and data management solutions that are perfect for AI workloads. As AI technology evolves, businesses must understand the infrastructure requirements for AI workloads in the cloud and employ the proper cloud computing solutions to maximize their AI workloads.

5 MACHINE LEARNING IN THE CLOUD

Machine learning is an artificial intelligence subfield that involves the use of statistical techniques to enable computer systems to learn from data and make predictions or judgments without being explicitly programmed. Because of its scalability, flexibility, and low cost, the cloud is a perfect platform for machine learning applications. In this chapter, we will go through how to apply machine learning on the cloud, covering data preprocessing, model construction and training, and model evaluation and optimization.

5.1 Data Preparation for Machine Learning in the Cloud

Preprocessing data is an important stage in machine learning that involves preparing the data so that the machine learning algorithm can use it. Some of the most important preprocessing approaches for cloud-based machine learning are:

1. **Data cleaning** is the process of eliminating or repairing any flaws or inconsistencies in data. Data cleansing is critical for maintaining machine learning models' accuracy and reliability.

2. **Data transformation**: Data transformation entails turning data into a format that machine learning algorithms can understand. Scaling, normalization, and feature engineering are examples of data transformation techniques.

3. **Data augmentation**: Data augmentation is the process of creating new data from existing data by using transformations like rotation, scaling, or flipping. Data augmentation can be used to expand the size of the training data set and improve the robustness of machine learning models.

5.2 Cloud-based Machine Learning Model Development and Training

After the data has been preprocessed, the machine learning models must be built and trained. Some of the most important strategies for developing and training machine learning models on the cloud are as follows:

1. **AutoML**: AutoML is a cloud-based solution that automates the creation and training of machine learning models. AutoML use machine learning techniques to develop and optimize models based on data.

2. **Distributed computing**: Distributed computing is breaking down a workload into smaller tasks and distributing them for processing over numerous servers or nodes. Distributed computing is useful for developing and training large-scale machine learning models.

3. **Cloud-based machine learning platforms**: Cloud-based machine learning platforms such as Amazon SageMaker, Google Cloud AI Platform, and Microsoft Azure Machine Learning offer a variety of tools and services for developing and training machine learning models.

5.3 Machine Learning Model Evaluation and Optimization in the Cloud

After the machine learning models have been built and trained, they must be evaluated and optimized. Some of the most important strategies for testing and optimizing machine learning models on the cloud are as follows:

1. **Model selection**: The best machine learning model is chosen based on performance indicators such as accuracy, precision, and recall. Model

selection is critical for assuring machine learning models' correctness and reliability.

2. **Hyperparameter tuning**: Hyperparameter tuning entails altering the machine learning algorithm's parameters to improve its performance. Tuning hyperparameters is critical for increasing the accuracy and reliability of machine learning models.

A/B testing compares the performance of multiple machine learning models or algorithms on the same data set. A/B testing can help you choose the optimum machine learning model or algorithm for a certain application.

Finally, because of its scalability, flexibility, and cost-effectiveness, the cloud is an ideal infrastructure for machine learning workloads. Preprocessing data is an important stage in machine learning that involves preparing the data so that the machine learning algorithm can use it. AutoML, distributed computing, and cloud-based machine learning platforms can be used to build and train machine learning models on the cloud. Model selection, hyperparameter tuning, and A/B testing are all strategies used to evaluate and optimize machine learning models in the cloud. As machine learning evolves, businesses must take advantage of cloud infrastructure to maximize their machine learning workloads.

5.4 Machine Learning Infrastructure in the Cloud

Businesses must have a dependable infrastructure that can manage the processing demands of machine learning workloads in order to construct and train machine learning models in the cloud. Some of the key infrastructure needs for cloud-based machine learning include:

1. **High-performance computing (HPC)**: HPC is the use of numerous processors and/or nodes to do computations at high rates. HPC is required for training machine learning models that need a significant amount of compute.

2. **Distributed computing**: Distributed computing is breaking down a workload into smaller tasks and distributing them for processing over numerous servers or nodes. Distributed computing is useful for developing and training large-scale machine learning models.

3. **GPU acceleration**: GPUs are specialized computers that can do parallel computations, making them suitable for training machine learning models. GPU acceleration is available from several cloud providers for machine learning applications.

5.5 Machine Learning Requires Cloud Storage and Data Management

Data is an essential component of machine learning, and enterprises must have a dependable storage and data management solution in place to support cloud-based machine learning workloads. Some major considerations for machine learning cloud storage and data management include:

1. **Data security**: Data security is critical for safeguarding sensitive data and adhering to data protection requirements. Businesses must ensure that their cloud storage and data management system has proper security measures in place, such as encryption and access control.

2. **Data accessibility**: For machine learning models to be effective, they must have access to enormous amounts of data. Businesses must ensure that their cloud storage and data management solution allows for quick and dependable data access.

3. **Data integration**: Machine learning models may require data from a variety of sources. Businesses must guarantee that their cloud storage and data management solution can integrate data from many sources.

5.6 Machine Learning in the Cloud: Challenges and Opportunities

While the cloud provides an ideal infrastructure for machine learning workloads, businesses must also consider some challenges and opportunities. The following are some of the most important difficulties and opportunities:

1. Cost: Building and training machine learning models in the cloud can be expensive, particularly for large-scale workloads. To ensure that machine learning on the cloud is cost-effective, businesses must carefully manage their costs.

2. **Skill shortage**: Building and training machine learning models necessitates highly specialized expertise. Businesses must ensure they have access to the skills and experience required to construct and train machine learning models in the cloud.

3. **Scalability**: The cloud infrastructure's scalability allows enterprises to quickly and easily grow their machine learning workloads. Businesses must, however, verify that their machine learning models are scalable and capable of handling rising workloads.

Machine learning in the cloud allows enterprises to take use of the scalability and flexibility of cloud infrastructure to quickly and effectively construct and train machine learning models. Businesses must, however, carefully examine the infrastructure needs, data storage and administration, and cost implications of cloud-based machine learning. Businesses can harness the full potential of machine learning in the cloud and drive innovation and growth with the right infrastructure and talent in place.

6 DEEP LEARNING IN THE CLOUD

Deep learning is an area of machine learning that focuses on the construction and training of multi-layered neural networks. From image identification to natural language processing, deep learning has transformed many industries and has the potential to impact many more. The cloud is a great platform for deploying and training deep learning models, allowing enterprises to take advantage of the cloud infrastructure's scale and flexibility. In this chapter, we will look at the principles of deep learning and neural networks, cloud deployment and training of deep learning models, and deep learning applications in various industries.

6.1 Deep Learning and Neural Network Fundamentals

Neural networks, which are made up of numerous layers of interconnected nodes, are used to create deep learning models. Each layer's nodes conduct a specified computation on the incoming data and send the result to the next layer. The layers closest to the input data are referred to as the input layers, while those closest to the output are referred to as the output layers. The layers in between are referred to as hidden layers. The model's architecture is defined by the number of hidden layers and the number of nodes in each layer. Deep learning models often include a large number of hidden layers and nodes, allowing them to learn complicated patterns in data.

Back propagation is used in neural networks to train the model. During training, the model is given a set of input data and the associated output data, which are referred to as labels. Based on the input data, the model generates a forecast, and the difference between the predicted and actual output is determined. This discrepancy is referred to as the loss. To minimize the loss, the model adjusts the weights of the nodes in each layer. This technique is performed numerous times until the model obtains an acceptable degree of accuracy.

• Deep learning models can use alternative optimization techniques, such as stochastic gradient descent or Adam optimization, in addition to back propagation.
• ReLU, sigmoid, and tanh are common activation functions used in deep learning.
• Convolutional neural networks (CNNs) are a type of neural network that is frequently employed for image recognition.
• Recurrent neural networks (RNNs) are a form of neural network that is frequently utilized in natural language processing jobs.

6.2 Deep Learning Model Deployment and Training in the Cloud

Deep learning model deployment and training in the cloud necessitates a dependable infrastructure capable of handling the processing demands of deep learning workloads. Some important factors to consider while deploying and training deep learning models in the cloud are:

1. **High-performance computing (HPC)**: Deep learning models necessitate a great amount of computation, which HPC can handle more efficiently. Many cloud providers provide high-performance computing (HPC) services for deep learning workloads.

2. **GPU acceleration**: GPUs are specialized computers that can do parallel computations, making them perfect for deep learning applications. GPU acceleration is available from several cloud providers for deep learning applications.

3. **Distributed computing**: Distributed computing is breaking down a workload into smaller tasks and distributing them for processing over numerous servers or nodes. Deep learning models that demand a substantial amount of compute benefit from distributed computing.

4. Cloud providers such as Amazon Web Services, Google Cloud, and Microsoft Azure provide a variety of services for deploying and training deep learning models, such as pre-built machine learning environments, GPU instances, and managed machine learning services.

5. Deep learning frameworks such as TensorFlow and PyTorch are extensively utilized on the cloud for developing and training deep learning models.

6. Deep learning frameworks for distributed deep learning, such as Horovod and Ray, can be utilized to distribute deep learning workloads across numerous cloud nodes or servers.

6.3 Deep Learning Applications across a Variety of Industries

Deep learning has a wide range of applications in a variety of industries, including:

1. **Healthcare:** Deep learning models can be used to analyze medical images, diagnose diseases, and develop new drugs. Deep learning models have been

utilized in healthcare to perform tasks such as skin cancer diagnosis, image analysis, and prediction of patient outcomes.

2. **Finance**: Deep learning models can be used to detect fraud, score credit, and anticipate stock market movements. Deep learning models have been applied in finance for tasks such as fraud detection, credit scoring, and stock market prediction.

Deep learning models in automotive can be utilized for autonomous driving, object detection, and driver behavior analysis. Deep learning models have been utilized in automotive for applications such as autonomous driving, object detection, and driver behavior analysis.

4. **Retail**: Deep learning models can be used to segment customers, recommend products, and estimate demand. Deep learning models have been applied in retail for tasks such as consumer segmentation, product selection, and demand forecasting.

Deep learning has already transformed numerous industries and has the potential to do so in many more. Deploying and training deep learning models in the cloud gives businesses the scalability and flexibility they need to realize deep learning's full potential. Businesses may design and train deep learning models that can learn complex patterns in data by grasping the principles of deep learning and neural networks. Businesses can unlock the full potential of deep learning in the cloud and drive innovation and growth with the right infrastructure and talent in place.

7 CLOUD-BASED AI PLATFORMS

Cloud-based AI platforms are a must-have for enterprises and organizations interested in leveraging artificial intelligence and machine learning technology. These platforms enable enterprises to design, train, and deploy machine learning models in the cloud by providing access to scalable computing resources and advanced machine learning algorithms. In this chapter, we'll look at the leading cloud-based AI platforms, compare their features and pricing methods, and talk about best practices and use cases for each.

7.1 Cloud-based AI Platforms Overview

Amazon Web Services (AWS), Microsoft Azure, and Google Cloud are the three primary cloud providers that provide AI systems. These platforms offer a wide range of cloud-based services for developing, training, and deploying machine learning models. Here's a quick rundown of each platform:

• **Amazon Web Services (AWS):** AWS provides a comprehensive set of machine learning services, such as Amazon SageMaker for developing and deploying machine learning models, Amazon Rekognition for image and video analysis, and Amazon Polly for text-to-speech conversion.

• **Microsoft Azure**: Azure provides a number of machine learning services, such as Azure Machine Learning for developing and deploying machine learning models, Azure Cognitive Services for natural language processing and computer vision, and Azure Databricks for big data and machine learning workloads.

• **Google Cloud**: Google Cloud provides a number of machine learning services, such as Google Cloud AI Platform for developing and deploying machine learning models, Google Cloud Vision for image analysis, and Google Cloud Natural Language for text analysis.

7.2 Features and Pricing Models Comparison

When choosing a cloud-based AI platform, it's critical to consider the platform's features and pricing models. Here's a comparison of AWS, Azure, and Google Cloud features and pricing models:

• **AWS**: For its machine learning services, AWS offers a pay-as-you-go pricing plan based on the amount of computation and storage resources needed. AWS also provides a selection of pre-built machine learning models that may be tailored to specific use cases.

• **Azure**: For its machine learning services, Azure has a consumption-based pricing model, with pricing based on the amount of compute and storage resources used. Azure also provides pre-built machine learning models, as well as a variety of third-party integrations.

• **Google Cloud**: For its machine learning services, Google Cloud offers a pay-as-you-go pricing plan based on the amount of computation and storage resources needed. In addition, Google Cloud provides pre-built machine learning models as well as a variety of third-party connectors.

When comparing features, consider the number of machine learning algorithms and frameworks supported, the platform's ease of use, and the level of automation provided.

7.3 Best Practices and Use Cases

Each cloud-based artificial intelligence platform has its own set of use cases and best practices. Here are a couple such examples:

• **Amazon Web Services (AWS):** AWS is a popular alternative for businesses wishing to build and deploy custom machine learning models. AWS SageMaker is a fully managed platform for constructing and training machine learning models, whereas Amazon Rekognition is a ready-to-use machine learning model for image and video analysis.

• **Azure:** Azure is an excellent alternative for businesses wishing to integrate machine learning into existing business operations. Azure Machine Learning offers a set of tools for developing and deploying machine learning models, whereas Azure Cognitive Services offers pre-built machine learning models for natural language processing and computer vision.

• **Google Cloud:** For enterprises wishing to exploit pre-built machine learning models, Google Cloud is a popular solution. Google Cloud AI Platform offers a variety of pre-built machine learning models, whilst Google Cloud Vision and Google Cloud Natural Language offer image and text analysis models, respectively.

When using cloud-based AI platforms, it is critical to adhere to best practices for security and data privacy.

AWS provides a variety of machine learning services, including Amazon SageMaker, Amazon Rekognition, Amazon Comprehend, and Amazon Polly. These services assist developers in creating, training, and deploying machine learning models on the cloud. Amazon SageMaker is a fully managed platform with pre-built algorithms and frameworks for constructing and deploying machine learning models. It also includes options for automatic model tuning, debugging, and monitoring.

Azure Machine Learning, Cognitive Services, and Bot Services are among the AI services provided by Microsoft Azure. Azure Machine Learning is a cloud-based platform for developing, implementing, and maintaining large-scale machine learning models. It is compatible with a wide range of programming languages and frameworks, including Python, R, and TensorFlow. Azure Cognitive Services delivers pre-built APIs for vision, speech, language, and decision making, making it simple for developers to add AI capabilities to their applications.

Google Cloud offers a variety of machine learning services, including Google Cloud AI Platform, Google Cloud AutoML, and Google Cloud Vision API. Google Cloud AI Platform is an all-in-one platform for developing and deploying machine learning models. It works with well-known machine learning frameworks such as TensorFlow, Keras, and PyTorch. Google Cloud AutoML enables customers to build bespoke machine learning models without any coding experience. The Google Cloud Vision API includes machine learning models that have been pre-trained for picture analysis and recognition.

There are various variables to consider when selecting a cloud-based AI platform, including affordability, ease of use, scalability, and security. It is also vital to assess the platform's degree of support and documentation, as well as the range of services and integrations available. Each platform has its own set of advantages and disadvantages, so it is critical to carefully assess each one to determine which is most suited to your requirements.

Cloud-based AI platforms can be employed in a variety of industries, including healthcare, banking, and retail, among others. In healthcare, for example, AI can be used to examine medical images and assist with diagnosis. AI may be used in finance to detect fraud and analyze risk. AI can be applied in retail to personalize shopping experiences and improve supply chain management.

When working with cloud-based AI platforms, it is critical to adhere to data security and privacy best practices. This involves ensuring that data is securely encrypted and that only authorized individuals have access. It is also critical to monitor and audit platform usage on a regular basis to verify compliance with legislation and policies.

Overall, cloud-based AI platforms provide a robust and adaptable platform for developing, deploying, and maintaining machine learning models on the cloud. Businesses may use these platforms to acquire insights and drive innovation by using the potential of AI.

8 CLOUD-BASED AI APPLICATIONS IN THE REAL WORLD

The use of cloud-based AI has transformed industries ranging from healthcare to banking to retail. In this chapter, we'll look at real-world applications of AI in the cloud, look at case studies of successful AI implementations, and discuss the ethical concerns and challenges of AI in the cloud.

8.1 Cloud-based AI Applications

Cloud-based AI has been employed in the **healthcare business** for a range of applications, including medical imaging, patient monitoring, and medication development. GE Healthcare, for example, has employed AI algorithms to examine medical pictures and assist with diagnosis. Furthermore, cloud-based AI has been used to remotely monitor patients, enabling healthcare professionals to provide timely care and improve patient outcomes. Furthermore, artificial intelligence (AI) has been used to accelerate drug discovery by analyzing large amounts of data to identify potential drug candidates.

Cloud-based AI has been employed in the **finance industry** for fraud detection, risk assessment, and customer support. American Express, for example, has utilized AI to detect fraudulent transactions in real time, saving millions of dollars. AI has also been used to assess credit risk, assisting financial organizations in making better loan decisions. Furthermore, artificial intelligence has been used to improve customer service by making personalized recommendations and responding to customer inquiries.

Cloud-based AI has been applied in the retail business **for inventory management**, supply chain efficiency, and customer experience. Walmart, for example, has employed artificial intelligence algorithms to optimize inventory levels, decreasing waste and increasing profitability. Furthermore, artificial intelligence has been used to improve supply chain visibility, allowing retailers to respond quickly to changes in demand. Furthermore, artificial intelligence (AI) has been used to personalize shopping experiences, providing customers with personalized recommendations and tailored promotions.

8.2 Case Studies of Cloud-based AI Deployments

Let's look at a case study of a successful cloud-based AI deployment to see how it works. British Airways debuted an AI-powered Chatbots in 2018 that was built with Microsoft's Azure Bot Service. "Flo," the Chatbots, was created to help clients with frequent questions and deliver personalized recommendations. Flo has handled over 50,000 talks per month since its inception, with a satisfaction rating of more than 70%. The Chatbots has aided in the improvement of customer service and the reduction of the workload on customer care employees.

However, as with any technology, there are ethical concerns and difficulties with AI in the cloud. Concerns have been raised concerning the possibility of bias in AI systems, which could lead to discriminatory consequences. Furthermore, there are concerns about data privacy and security because sensitive data is frequently stored in the cloud. Organizations must examine these ethical considerations and put protections in place to mitigate these dangers.

Finally, cloud-based AI has transformed a variety of businesses by providing strong tools for data analysis, decision-making, and automation. Businesses may acquire useful insights, boost operational efficiency, and improve consumer experiences by adopting cloud-based AI. However, organizations must carefully consider the ethical concerns and challenges associated with AI in the cloud and implement appropriate safeguards to ensure responsible use of this technology.

Cloud-based AI applications in healthcare could include medical image analysis, patient monitoring, and drug discovery. AI applications in banking could range from fraud detection to portfolio management. AI can be applied in the retail industry for tailored product suggestions, supply chain optimization, and customer support Chatbots.

Case studies of successful cloud AI implementations could include Netflix, which utilizes machine learning algorithms to personalize movie and TV program suggestions, and Airbnb, which employs AI to optimize pricing and search results for their consumers.

The ethical implications and problems of AI in the cloud could include subjects such as privacy concerns, prejudice in AI algorithms, and the possibility of AI replacing human jobs. It might also go over the necessity of openness and accountability in AI systems, as well as ensuring that AI is used for the good of society as a whole.

9 AI CLOUD SECURITY AND GOVERNANCE

As more businesses implement cloud-based artificial intelligence (AI) solutions, it is critical to emphasize security and governance in order to protect sensitive data and comply with data protection rules. In this chapter, we will look at best practices for cloud security and AI governance, such as data privacy and compliance legislation, risk management techniques, and security threat mitigation.

9.1 AI Cloud Security Best Practices:
• To manage and monitor security across all cloud services, consider leveraging cloud-native security services such as AWS Security Hub or Azure Security Center.
• Use secure coding methods to avoid typical vulnerabilities like injection attacks and cross-site scripting (XSS).
• Create disaster recovery and business continuity plans to ensure that essential systems are rapidly restored in the event of a power outage or disaster.
• Monitor and restrict privileged account access to avoid unauthorized access to sensitive data or systems.
• Use network segmentation to isolate and restrict access to sensitive data.

9.2 Cloud-based AI Data Privacy and Compliance Regulations:
• Be familiar with and follow data privacy standards such as the General Data Protection Regulation (GDPR), the California Consumer Privacy Act (CCPA), and the Health Insurance Portability and Accountability Act (HIPAA).

• Ensure that data is stored and processed in accordance with data protection legislation in force.
• Develop data management rules and processes, such as data classification, access limits, and retention.

9.3 Risk Management and Security Threat Mitigation Strategies:
• Create a risk management strategy that includes risk identification, mitigation, and monitoring.
• Put in place a disaster recovery plan to mitigate the effect of data breaches or system failures.
• Conduct frequent security audits and penetration testing to discover and remedy issues.
• Stay current on the latest security threats and use best practices for incident response.

Conclusion: Cloud security and governance are essential components of an effective AI strategy. Organizations may protect sensitive data and enable the effective implementation of AI solutions in the cloud by establishing best practices for cloud security, ensuring compliance with data privacy legislation, managing risk, and minimizing security risks.

10 CLOUD COMPUTING AND AI IN THE FUTURE

Cloud computing and artificial intelligence (AI) are rapidly emerging technologies that continue to drive innovation across multiple industries. In this chapter, we will look at developing trends and technologies in cloud computing and AI, as well as their possible repercussions on society and the workforce, as well as projections for their future.

10.1 Cloud Computing and AI Emerging Trends and Technologies
Every year, new trends and technologies emerge in the cloud computing and artificial intelligence industries. Among the most prominent upcoming

concepts and technologies in cloud computing and artificial intelligence are:

- **Multi-cloud environments**: As more businesses adopt cloud computing, many are opting for multi-cloud settings, which allow them to use different cloud platforms for different applications and workloads. The desire for increased flexibility, improved performance, and lower costs is driving this trend.

- **Hybrid cloud architectures**: Hybrid cloud designs are gaining popularity because they enable enterprises to benefit from both public and private cloud environments. Businesses can keep sensitive data on-premises while enjoying the scalability and flexibility of the public cloud by implementing a hybrid cloud architecture.

- **Edge computing**: An edge computing paradigm involves processing data closer to the source of the data rather than in a central data center or cloud. This method is appropriate for IoT devices and autonomous vehicles that require real-time processing or low-latency connectivity.

- **Quantum computing**: Quantum computing is a new technology that has the potential to change the way we process and interpret data. Quantum computers do calculations using quantum bits (qubits), which can solve some problems tenfold quicker than ordinary computers.

The Effects of Cloud Computing and Artificial Intelligence on Society and the Workforce

Cloud computing and AI technologies are having a big impact on society and the workforce as they evolve. Among the most significant effects are:

- **Employment automation**: As AI technologies improve, there is rising concern that many positions will be automated, resulting in major employment losses across multiple industries.

- **Improved efficiency and productivity**: By automating repetitive processes, minimizing errors, and offering real-time insights into corporate operations, cloud computing and AI can help firms enhance efficiency and productivity.

- **improved healthcare outcomes**: AI technologies can assist healthcare providers in improving patient outcomes by offering more accurate diagnoses, individualized treatment plans, and improved patient health monitoring.

- **Environmental sustainability**: By optimizing energy consumption, eliminating waste, and supporting sustainable business practices, cloud computing and AI can help firms lessen their environmental impact.

10.2 Cloud Computing and AI Predictions for the Future

Experts anticipate that as cloud computing and AI advance, they will have a dramatic impact on numerous industries and society as a whole. Among the most important forecasts regarding the future of cloud computing and AI are:

• **True AI consciousness**: Some experts believe that AI systems will someday become sentient, resulting in true AI consciousness. This might have far-reaching consequences for how we interact with machines and the environment around us.

• **New computing paradigms**: As data grows at an exponential rate, new computing paradigms will be required to handle the vast amounts of data produced by IoT devices, social media platforms, and other sources.

• **Healthcare advancements**: AI technology will continue to evolve, resulting in more accurate diagnoses, individualized treatment regimens, and improved patient outcomes.

• **Increased automation**: Automation will remain a major trend in the workforce, with more occupations being automated as AI technologies progress.

• **Increased cybersecurity threats**: As cloud computing and AI grow more prevalent, the risk of cybersecurity threats increases, necessitating the need for stronger cybersecurity solutions.

Finally, cloud computing and artificial intelligence (AI) technologies are continuously changing, with new trends and technologies appearing every year. As these technologies advance, they will have a significant impact on a variety of industries.

ABOUT THE AUTHOR

Isaac Odun-Ayo is a Professor of Computer Science at Chrisland University, Abeokuta, Nigeria. He has several conference and journal publications in Cloud Computing. He has several eBooks on cloud computing. He also has an IBM Cloud Application Developer Mastery Award. He loves reading, writing and research. His hobbies include traveling and playing golf. He is married and with children.

Books by This Author

Cloud Computing Architecture and Service Models

Cloud Virtualization and Containers

Cloud Storage and Data Management

Cloud Security and Compliance

Cloud Networking and Interconnectivity

Cloud Automation and Orchestration

Cloud Deployment and Scalability

Cloud Monitoring, Analytics, and Optimization

Cloud Migration Strategies and Best Practices

Cloud Application Portability

Cloud Multi-Tenancy

Cloud Computing Basics

Basics of Cloud Computing and Machine Learning